REPORT

OF THE

INSPECTION OF THE SOUTH PASS IMPROVEMENT

OF THE

MOUTH OF THE MISSISSIPPI RIVER,

(JUNE 9, 1876,)

BY

C. B. COMSTOCK,

MAJOR OF ENGINEERS, BVT. BRIG. GEN., U. S. A.

WASHINGTON:
GOVERNMENT PRINTING OFFICE.
1876.

44TH CONGRESS, } SENATE. { EX. DOC.
1st Session. } { No. 77.

LETTER

FROM

THE SECRETARY OF WAR,

TRANSMITTING

A report of Major Cyrus B. Comstock, Corps of Engineers, of his inspection of the South Pass improvement of the mouth of the Mississippi River.

JUNE 20, 1876.—Referred to the Committee on Commerce and ordered to be printed.

WAR DEPARTMENT, *June* 19, 1876.

The Secretary of War has the honor to transmit to the United States Senate and House of Representatives report and maps of the last inspection made by Major Cyrus B. Comstock, Corps of Engineers, United States Army, of the South Pass improvement of the Mississippi River, with copy of the instructions pertaining to the same issued to him from the office of the Chief of Engineers, dated the 25th of April last.

J. D. CAMERON,
Secretary of War.

OFFICE OF THE CHIEF OF ENGINEERS,
Washington, D. C., June 17, 1876.

SIR: I transmit herewith Major C. B. Comstock's report and maps of his last inspection of the South Pass improvement of the mouth of the Mississippi River, prefixing to it his instructions from this office of the 25th April last, previous to the inspection.

Very respectfully, your obedient servant,

A. A. HUMPHREYS,
Brigadier-General and Chief of Engineers.

Hon. J. D. CAMERON,
Secretary of War.

LETTER OF INSTRUCTIONS.

OFFICE OF THE CHIEF OF ENGINEERS,
Washington, D. C., April 25, 1876.

SIR: It has become desirable that in your next inspection of the South Pass improvement you report fully all the information, with your judgment thereon, that the Secretary of War is required, in paragraph

10, section IV, of the river and harbor act of March 3, 1875, to report to Congress during the construction of the jetties and auxiliary works; especially the material used, the character and permanency of the jetties and auxiliary works, and whether they are so thoroughly substantial and permanent as to maintain the improved channel for all time after their completion.

Further, if the jetties and other works do not possess the degree of strength and durability specified in the act of Congress, you will please state to what degree they fall short of it, and further report to what degree the jetties and auxiliary works are less massive, less strong, less enduring, and less costly than the works planned by the Board of Engineers, constituted under the river and harbor act of June 23, 1874, of which Lieutenant-Colonel Wright was president and yourself a member, the report and plan of which Board was submitted to Congress, and was before that body when the river and harbor act of March 3, 1875, was passed, and the contract with Mr. Eads entered into.

In the inspection and report you are now desired to make you will include the channel and works at the head of the South Pass, and treat in the same manner all particulars relating to them.

You will also cause soundings to be made with considerable detail for a distance of half a mile, or thereabout, just outside of the bar, and have soundings also carried in less detail for some distance beyond.

Very respectfully, your obedient servant,

A. A. HUMPHREYS,
Brigadier-General and Chief of Engineers.

Maj. C. B. Comstock,
Corps of Engineers.

REPORT.

Office United States Lake Survey,
Detroit, Mich., June 9, 1876.

General: I have the honor to report the condition of the works for the improvement of the South Pass of the Mississippi River on April 30, 1876.

The soundings on the three sheets of tracings transmitted herewith are of dates from April 27 to May 12. All soundings referred to in this report, except those in Grand Bayou, whether made by the Coast Survey or under my direction, are reduced to the plane of average flood tide.

The location of the jetties and the method of constructing the mattresses has been fully described in my previous reports. As there has been little change in either, it is unnecessary to repeat the description.

EAST JETTY.

The work done on this jetty since February 7, 1876, has been confined to raising the jetty in low places by additional mattresses, and to extending its mattress-work toward the point called the end of jetty, which is 11,941 feet from beginning of jetty, or 12,100 feet from station East Point. On April 30, this extension had been brought up or nearly up to the water-surface, to within 400 feet of the end of jetty. Along the line of sheet-piling, which continues for about 5,000 feet from the beginning of the jetty, additional mattresses have been placed in the second tier, so that now there are two tiers of mattresses for nearly the whole length of the sheet-piling.

In general, it may be said that the east jetty now confines the water at low tide pretty thoroughly to within 400 feet of the end of the jetty. There are, however, places where the sheet-piling has washed out, or where there are holes through the mattress-dike, which permit small amounts of water to pass through the east jetty, thus giving shallow channels through the general shoal which exists outside of this jetty. There has been some shoaling on the outside of this jetty since the work was commenced.

Starting from East Point and going seaward on a line parallel to and 100 feet outside of east jetty, the following table gives for each 2,000 feet the average depth, as shown by the Coast Survey map of May and June, 1875, and by the map of soundings forwarded herewith, of May, 1876, both being referred to average flood-tide.

	Distances in feet.					
	0–2,000.	2,000–4,000	4,000–6,000	6,000–8,000	8,000–10,000	10,000–11,000
	Feet.	*Feet.*	*Feet.*	*Feet.*	*Feet.*	*Feet.*
Depths, 1875	3.4	3.2	3.2	4.0	4.0	6.5
Depths, 1876	1.0	1.2	1.4	3.0	4.0	3.5

Within a few hundred feet of the end of the east jetty, piles have disappeared from time to time. In some cases they may have been washed up by heavy seas, or the teredo may have injured some enough to enable the sea to break them off; but the evidence seems to show that many have sunk gradually out of sight, although their tops were at first several feet above water. At this place the top of the jetty has also sunk, and, apparently, not from its top mattresses being destroyed by the waves. Both seem to indicate a settling of the strata into which the piles are driven, or upon which the mattresses rest, to a distance of from 5 to 10 feet. Less marked phenomena of the same kind have been observed in the west jetty; in places where the settling of the top of the jetty is small, it might be attributed to the compression of the lower mattresses.

The cross-section of the jetties remains essentially the same as heretofore. The outer edge of each mattress rests against the vertical piles. The bottom mattresses vary from 35 to 50 feet in width; where the water is so deep as to require from 3 to 6 layers, the top mattresses are from 22 to 25 feet in width. An ordinary mattress occupies about 3 feet in height in the jetty. Details of the state of the mattress-work are shown on sheet No. 2, herewith.

WEST JETTY.

As in the east jetty, the work on the west jetty since February 7, 1876, has been mainly confined to the filling of gaps in the jetty where it had not been brought up to the water-surface, and to the extension of its mattress-work.

On April 30, 1876, the mattresses were up, or nearly up, to the low-water surface from the head of the jetty to within 200 feet of pile No. 355, which is opposite the end of east jetty, and has heretofore been called the end of the west jetty. The number of layers of mattresses along this jetty is from three to seven, varying with the depth of water and the amount of settling. The advance of the outer face of the bar, which will be referred to hereafter, gives now but 11½ feet of water at pile No. 355, where a year ago the Coast Survey found 21 feet, and the piling for the jetty has accordingly been extended about 250 feet beyond pile No. 355.

Extensive shoaling has taken place on the outside of the west jetty, as appears from the following table, which gives, at average flood-tide, the depths, 200 feet outside of west jetty, for each 2,000 feet below East Point station, derived from the Coast Survey map of June, 1875, and from the recent survey:

	Distances in feet.			
	4,000-6,000.	6,000-8,000.	8,000-10,000.	10,000-12,000.
	Feet.	*Feet.*	*Feet.*	*Feet.*
Depths, May and June, 1875	14.1	13	9.8	7
Depths, May, 1876	6	2	1.8	1.5

The shoaling can be seen on the sections across jetties given on sheet No. 2 of tracings, herewith, showing the progress of mattress-work and giving details.

It will be noticed that in 1875 shoals existed a little to the east of the east jetty, and west of west jetty for nearly their whole lengths, and that much of the subsequent shoaling outside the jetties has been in the deeper water between the crests of these shoals and the lines of the jetties.

SCOUR.

At 8,210 feet below East Point station a line of piles 150 feet long has been driven, starting from east jetty, and running at right angles to it toward the middle of the pass. A similar opposite line projects from the west jetty. It seems to have been the intention to have put aprons against these spurs of piling, in order to obtain greater scour between them, but the plan has not yet been carried out.

Similar spurs of piling were driven at 9,450 feet below East Point, but the spur-work was not completed.

Some attempts have been made at stirring up the bottom, and thus aiding scour, both at the head and at the mouth of South Pass.

A long chain, with a heavy pulley at its end, was dragged over portions of the bottom, but this process was soon abandoned. A kind of rake with iron teeth, which had two arms running from the ends of its head past the sides of a scow, to a transverse axis on the scow, about which the rake revolved, was also tried; but this was broken while the scow was being towed, dragging the rake. At my recent visit a stronger one had just been constructed and was in operation, but I do not yet know with what results.

Upon comparing the Coast Survey map of June, 1875, with sheet No. 1, herewith, both being referred to the plane of average flood-tide, it will be seen that at the first date the lowest point of the 30-foot curve was 300 feet below East Point, where there was a hole of that depth, which has now increased a little in size, but that there has been no marked change in the pass from the head of the east jetty to near the head of the west jetty, about 4,000 feet below. In June, 1875, 18¾ feet could be carried down to the head of the west jetty, now 20¼ feet can be so carried. It will be seen that the 20-foot curve then had its lowest point 2,600 feet below East Point, and that now the lowest point of the main 20-foot curve is 8,900 feet below East Point. This 20-foot channel has an average width of about 200 feet from its lower end, where it is very narrow, to the head of west jetty, a distance of 4,900 feet. The distance from its lower end to 20 feet of water outside of bar is now 3,300 feet.

Perhaps a clearer statement would be the following table, showing the draught of water that could be taken through each 2,000 feet of the channel below East Point station in June, 1875, and in May, 1876:

	Distances in feet.					
	0–2,000	2,000–4,000	4,000–6,000	6,000–8,000	8,000–10,000	10,000–12,000
	Feet.	*Feet.*	*Feet.*	*Feet.*	*Feet.*	*Feet.*
Draught, 1875.	22.5	18.7	16.7	10.2	9.7	9.2
Draught, 1876.	23.3	20.3	22.0	21.0	17.1	15.7

At the ends of the jetties the bar seems subject to considerable changes in short periods. In the soundings of May 3 Captain Brown found a least depth in channel of 15 feet. On May 10 and 11 I ran numerous lines of soundings across the bar, but failed to find more than 13½ feet on any line. I was informed that between Captain Brown's soundings and mine a strong southeast wind had blown for several days, which may have caused the shoaling. A few additional soundings on May 28, by Mr. Parmely, indicated that, from scour and raking, the depth on the bar had again increased to 16½ feet.

In considering the advance of the outer face of the bar at the ends of the jetties, the movement of the 20-foot curve will be a good guide. In June, 1875, this curve, gently convex toward the sea, crossed the present line of the east jetty about 300 feet above its end, (pile 1,101,) and the present line of the west jetty, about 20 feet above pile No. 355, hitherto called the end of west jetty. On sheet No. 1 it will be seen that now the 20-foot curve has advanced on the line of the east jetty about 250 feet, and on the west jetty about 160 feet, while on a line half-way between the two jetties it has advanced about 470 feet. Its average advance in front of the jetties may be taken at 350 feet. If the 30-foot curve be examined the same movement is observed, but to a smaller amount, say about 200 feet on an average. This curve has moved very irregularly, some points receding instead of advancing.

In 1875 the Coast Survey map shows that on passing the outer crest of the bar, where the ends of the jetties now are, the water deepened abruptly to 35 to 45 feet, and then shoaled. On this shoal, which ran obliquely in front of the jetties, and at a distance of about 800 feet from their ends, there was, except in a narrow channel through it, less than 30 feet of water, the minimum depth being 13¾ feet. At present the narrow channel has disappeared; and limiting the word "shoal" to the area which has less than 30 feet of water on it, it may be said that the shoal appears to have moved from 200 to 300 feet seaward, retaining about the same area and the deeper water between it and the ends of the jetties, and that it has now a minimum depth on it of 14½ feet. Going outside this shoal, the 40-foot curve, while irregularly changed during the year, yet shows little or no general advance.

The shoaling which has occurred is shown on sheet No. 3, herewith, which gives three parallel sections of the gulf bottom, running seaward from the ends of the jetties—the first in the prolongation of the east jetty, the second parallel to the first and half-way between the jetties, and the third parallel to the others, and from pile 355, or end of west jetty.

The sections are constructed from Mr. Marindin's survey of May and June, 1875, and from Captain Brown's survey of May, 1876, both reduced to average flood-tide.

These sections show in detail for these three lines, out to the 50-foot curve, what has already been stated in a general way for the whole area in front of the ends of the jetties.

The eastern section indicates very irregular changes in the first 1,500 eet, showing deepening in some places and shoaling in others, without a decided preponderance of either, except near the end of the east jetty.

The central section shows the shoaling at the end of the jetties and the apparent advance of the shoal 1,000 feet in front, already referred to.

The western section shows shoaling near the end of the west jetty.

All these sections indicate a considerable deepening at distances in front of the jetties, between 2,400 and 4,000 feet.

In a material so soft as the mud deposited in deep water it is difficult to say what irregular changes may arise, either from its settling or from wave-action on it in storms. But a deepening here of eight or ten feet does not, in advance, seem probable, and the work will at once be repeated, to detect the error or verify the results of the recent survey of this area.

GRAND BAYOU.

The work at the head of Grand Bayou, for its closure, has been suspended, apparently from the difficulty in building a dam across it there, and its closure has been undertaken at a point about a mile down the pass. Three tiers of mattresses, each 3½ feet in thickness, have been sunk across the bayou; the mattresses of the lower tier being 100 feet long, their axes being in the direction of the stream. The top layer of mattresses has from 11 to 24 feet of water on it. The state of the work is shown on sheet No. 1.

HEAD OF SOUTH PASS.

The position of the jetty was described in my last report, and is shown on sheet No. 1, herewith, as is also the condition of the mattress-work. Four tiers of mattresses have been sunk for about two-thirds the length of the jetty, on the west side of the guide-piles, and in places the number of tiers reaches six. On the east side of the guide-piles a single tier of mattresses has been sunk for four-fifths the length of the jetty.

At the present high stage of the river much of it is not up to the river-surface, and the water at all such places runs across it at an angle of 20 or 30 degrees, with high velocity, the obstruction of the work giving a small head.

Details of the work are shown on sheet No. 1. If a line parallel to the jetty and 1,000 feet west of it be drawn from the head of the island to a point opposite the upper end of the jetty, a comparison of Mr. Marindin's survey of June, 1875, with Captain Brown's survey of May, 1876, shows a great shoaling in this area of 2,600 feet by 1,000 feet. In 1875, 14 feet of water could be carried through this area into South Pass; now but 9 feet can be taken.

Then the average depth in the upper third of the area described was about 17 feet; now it is about 10 feet.

For the middle third this average depth has changed from 14.5 feet to 8.6 feet.

For the lower third it has changed from 14.6 feet to 9.5 feet. On comparing the soundings in the west entrance to South Pass, on sheet No. 1, herewith, with those of Mr. Marindin's survey of June, 1875, no very marked change appears; certainly no scour comparable with the shoaling in the east entrance.

It is not always easy to assign the efficient cause for shoaling in a special portion of the river. The shoaling in the east entrance may have arisen from some general causes, which would have acted had no attempt been made to improve the river. But the changes which have been made in the South Pass might produce shoaling—

1st. At the lower end of the jetties at mouth of South Pass the velocity is much greater than it was a year ago.

This increased velocity requires an increased head to produce it, and the surface of the pass between the jetties has, at least in high-river stages, become elevated. Grand Bayou has been partially obstructed, raising its water-surface. It is probable that both these causes raise the water-level somewhat in the South Pass at its head, thus diminishing the velocity of the water entering it from the main river. If the entering velocity is sufficiently diminished, some of the sediment carried must be dropped.

I have no precise data for estimating the rise in the water-surface between the jetties, produced by their construction. Some months since there seemed to be a difference of eight or ten inches in water-level on the inside and outside of Kipp dam.

If we suppose the water-level of the pass at that place to have been raised by six inches above its original position, hydraulic formulæ (which are, however, confessedly based on imperfect theories, but which are the best we have) give a resulting rise of about three inches at head of South Pass when the stage of the river is a medium one, giving a slope of one foot in this ten and a half miles distance, the pass being supposed of indefinite length.

2d. The jetty at the head of the South Pass, where not raised to the water-surface, is now crossed at an angle of 20 or 30 degrees by water running with considerable velocity into Pass à l'Outre.

It acts, therefore, to some extent as a dam, diminishing the velocity of the water above it, and perhaps causing it to deposit a part of its sediment.

Which of these causes has been most effective in producing shoaling I am unable to say.

On April 25, 1876, you gave me instructions to report "the material used, the character and permanency of the jetties and auxiliary works, and whether they are so thoroughly substantial and permanent as to maintain the improved channel for all time after their completion;" also to state "to what degree they fall short of it, and further report to what degree the jetties and auxiliary works are less massive, less strong, less enduring, and less costly than the works planned by the Board of Engineers constituted under the river and harbor act of June 23, 1874, of which Lieutenant-Colonel Wright was president, and yourself a member, the report and plan of which Board was submitted to Congress and was before that body when the river and harbor act of March 3, 1875, was passed, and the contract with Mr. Eads entered into."

The first question is whether the jetties and auxiliary works are so "substantial and permanent as to maintain the improved channel for all time after their completion." It includes the questions whether the works are in themselves permanent, and whether they will maintain a permanent deep channel.

At present the jetties at South Pass are embankments of superposed mattresses, one side of the embankment being vertical, the other side so sloping that while the width of the jetty at the water-surface is 25 feet, the width at bed of river is from 35 to 50 feet. The vertical side of the embankment is in contact with a row of piles; where the wave-action is strong, piles have also been driven along the other edge of the top mattress, to prevent its displacement by waves. A part of the mattresses have been sunk by rubble-stone thrown upon them, a part by allowing mud to accumulate in them while afloat over their proper positions. In their present condition these jetties are not permanent structures. As Mr. Eads has not furnished me with copies of his plans, I am unable to state definitely what further work he proposes to complete them, but understand it is intended to widen and cover them with stone.

As to whether, if these works are made permanent, they will with moderate occasional extensions maintain the improved channel for all time after their completion, I would state that the information on which judgment is to be based is, with some additions, the same as that before the Board of Engineers (established under the act of Congress of June 23, 1874) which proposed the improvement of the South Pass, and first gave a definite plan for the work.

The additional information on which I place reliance is that already referred to in considering scour, derived from comparison of the Coast Survey map of June, 1875, and sheet No. 1, transmitted herewith. The comparison shows an average advance in front of the jetties of the outer face of the bar, as defined by the 20-foot curve, of about 350 feet.

It seems probable that a part of the deposit producing this advance is that scoured from between the jetties and rolled along the bottom, or carried in intermitting suspension near it, to be dropped as soon as it comes to deeper and stiller water outside of the bar and below the level of its crest. An acceleration in the advance of the bar would naturally be expected while this scour was going on, and will continue so long as the scour continues; when that scour ceases the advance of the bar will be independent of this temporary and extraordinary cause of advance.

The present rate of advance, arising in part at least from scour, gives, in my judgment, no definite information as to what the ultimate rate will be, and hence does not give sufficient data to modify the opinion expressed by the Board of Engineers, already referred to, as to the permanence of the channel when properly jettied, and with moderate extensions of the jetties from time to time.

In the second place I am directed to report "to what degree the jetties and auxiliary works are less massive, less strong, less enduring, and less costly than the work planned by the Board of Engineers" already referred to.

The Board proposed for the first 7,100 feet of the east jetty a mattress-dike 10 feet wide on top, with side slopes of $\frac{3}{8}$. In its present condition the first 5,000 feet consists of two rows of piles, a line of sheet-piling, and two layers of mattresses. The next 2,100 feet has a cross-sectional mattress area somewhere near that proposed by the Board. From this point on, the jetty in its present state has a less cross section than that proposed by the Board, the difference becoming very great on reaching deep water, where the Board proposed a top width of 50 feet, with side slopes of $\frac{1}{4}$, while the construction has a top width of 25 feet and bottom width of 50 feet.

The head of the west jetty, as constructed, is about 4,000 feet below that of east jetty. The same relations exist between it and the plan of the Board as in the corresponding portion of the east jetty.

The Board proposed mattresses essentially the same as those used in the only similar mattress-structure which has successfully resisted heavy seas, consisting of crossed layers of strong fascines, whose intervals were filled with brush. The mattresses used at South Pass have no fascines, but are layers of leafy brush, confined by strips of boards on top and bottom, and by stakes, which pass through the mattresses, whose heads are fastened into these strips. The cost of the mattresses proposed by the Board would have been from 50 to 100 per cent. greater per cubic foot.

Shoals, previously existing or newly formed, so protect the outside of both jetties that they are not now exposed to violent wave-action, except toward their sea-ends. For the protected portions the mattresses used may very probably prove to have sufficient strength when covered by stone.

The Board proposed a substantial covering of stone for the jetties. No such covering has as yet been applied to any part of them.

The dike at the head of South Pass is constructed in the same way as those at the mouth; is as yet incomplete, and has no stone covering.

In their present state the jetties have not nearly reached the massiveness proposed by the Board of Engineers. They have no stone covering, so that they are less enduring, and the mattresses are much more cheaply constructed than those proposed.

Very respectfully, your obedient servant,

C. B. COMSTOCK,
Major of Engineers and Brevet Brigadier-General, U. S. A.

Brig. Gen. A. A. HUMPHREYS,
Chief of Engineers, U. S. A.

www.ingramcontent.com/pod-product-compliance
Lightning Source LLC
LaVergne TN
LVHW020643110826
845149LV00004B/1324
* 9 7 8 1 4 1 8 1 8 9 6 5 5 *